taf^oni

NATURAL DESIGN OF WEATHERED STONE

CRISTINA I. VIVIANI
author

MAUREEN WELTON
creative director

KENNETH TORRANCE
design

18KARAT

taf^{o}ni

misteriose forme scolpite dalla natura

nature's mysterious sculptural forms

00.004.005

There's a place I go to reconnect, to reclaim myself, to be reminded of that which is constant and that which is simply transitory. This is a place on the ocean, an eroded sandstone beach, where it seems that life begins on the boundaries, between water and land, heaven and earth, the real and the imagined. ¶ On this beach the insistent wind can chill me to the bone; its cold fingers reach beneath each layer of clothing. Stormy weather with its accomplice, melancholy, can hang heavily for days. Although some weather is gentle and benign, the fresh, sunny days filled with brilliant light are what I most anticipate. When I hear the excited voices of children discovering yet another purple starfish, I breathe in a moment of grace. ¶ I walk as far as I can and allow my mind to set aside whatever it has been struggling with. The natural world embraces me. The carved and eroded sandstone cliffs provoke my curiosity; I question their existence, their formation, the geology and chemistry of their creation. I marvel at the intricate patterns and textures, the forms, the in-between spaces, the layers of hardened sediment. What causes all of this? As the light plays with eroded rocks, different shapes appear: humanistic faces, animals, reflections of the animate within the inanimate. When the light is low they feel unsettling – I feel stone faces watching me, examining me, as I walk along the shore. ¶ When I return, I feel much lighter. My mind's weight has been released, realigned, and I know that everything will be fine. All is just as it should be. These unstructured encounters with nature release the non-essential and revive my creative spirit. Is this a release of latent creativity stored in the caverns of the brain, or is it new energy that enters from another consciousness? Whatever the source, this creative energy starts to dance and new, expansive thoughts race through my body clamouring for attention. I feel alive again. ¶ This book is about that place where I go – where many people go – that special place where we reclaim our essential self. It is a creative exploration of the peculiar weathering patterns and circular erosion of the stone cliffs and beaches, called tafoni, found in the Gulf Islands of British Columbia, Canada. We explore the scientific formation of the natural sculptures and use tafoni as a metaphor for the dramas and transitions of human life. It is the source of inspiration for the design of 18Karat products that are handmade in places around the world; we study the textures, patterns, shapes, and moods, translating them into objects we want to have in our surroundings. It's also the impetus for a sensory experience in our store. *Tafoni* is our collective expression, a collaboration between creative people, a synthesis of our knowledge, personal experiences and emotions. Cristina researched the geology and science of the formations, brought us the name, *Tafoni*, took most of the photographs, and wrote the text and poetic reflections. She lived this book for almost a year and, in a way, this is her story. Kenny breathed life into the book and created its graphic expression. I was the metaphorical glue, the one who envisioned and held the parts together and encouraged its existence. It is our wish that this book takes you to that place where the noise of daily living is replaced with the beauty and wisdom of the earth, to that place where these quiet words can live inside of you and inspire your own creativity, at least until you can be on your own beach again.

Maureen Welton

afoni is the Italian word for eroded caverns and rounded hollows tha orm on cliffs, boulders and shorelines. It is the name that geologists have given to clusters of mysterious rock cavities that often resemble organic, sculptural forms and curious symbols. Tafoni are said to *grow* on crystalline and granular rocks and seem to form in places around the world where more extreme environmental conditions exist. ¶ The underlying geology and climate of the Gulf Islands of British Columbia provide the ideal setting for exquisite tafoni formations. Millions of years ago, the Earth's tectonic plates suddenly altered their course and thick layers of marine sediments along the east side of Vancouver Island buckled and folded. Tilted outcrops of ancient geological formations surfaced along the west side of the Georgia Strait. Today, more than 200 islands – some barely exposed rocks, others quite mountainous – appear crinkled under the sun, wind, rain, and rising tides. Over time some of the shorelines on the islands have weathered and eroded to form small havens of sensuous tafoni ridges and hollows. ¶ *Tafoni* is a book for anyone who has ever been inspired by weathered rocks – by the raw beauty of nature's mysterious sculptural forms. Although there have been numerous studies of tafoni, their exact formation process remains elusive and little has been published about these particula weathered rocks. We wanted to create a book that would reflect the beauty of tafoni, not with the intention of writing a book about geology but rather to publish an informative and inspirational book about tafoni For those who wish to delve further into the subject we have included a glossary, further readings and a list of the more accessible places on the Gulf Islands where tafoni and other unusual weathered rocks can be found. ¶ To explore these puzzling and somehow familiar forms which reveal a haunting and intimate face of the Earth, I chose to write this book in two distinct voices: one that is more formally structured and contains relevant geological information and the other in the form of poetic reflections, which are based on personal observations experiences and speculations. I wanted to bring attention to one of the lesser known mysteries on Earth, one that emanates from a fluid movement of cycles and change. Tafoni and weathered rocks are reminders of the resilience and tranquility residing deep within us particularly during times of uncertainty and change. **Cristina I. Vivian**

Earth contains minerals, liquids and gases gathered from the Milky Way. It formed from drifting interstellar clouds travelling through space, 4.6 billion years ago. Metallic minerals converged and fused at the core; lighter, more crystalline rocks floated to the surface, 4.55 billion years ago. ¶ Massive meteorites struck the Earth for the first 100 million years of its existence – several times melting it to the core. Million tonne rocks collided with the Earth until about 3.9 billion years ago. 50 million years later sophisticated microbes were well-established in a hot, volatile and acidic environment. ¶ Emerging from a fiery core, the Earth brews well-kept secrets and stories, revealed mostly during times of unrest. Mysterious and sensuous, tafoni sculptures and weathered rocks appear out of the Earth's ancient and fluid evolution.

Tafoni and weathered rocks

unveil the magnificence of

the Earth's circling, fluid existence,

the fathomless eternity of time

and the inescapability of change.

This spiraling force

pervades our whole existence.

Our sun revolves around the Milky Way.

The Earth orbits the sun,

the moon circles the Earth,

ocean currents, wind and water spiral,

yet, we remain relatively stable

on a small, semi-molten,

spinning mass.

if the earth turned **45 years** old today

homo erectus arrived about **06 days** ago

homo sapiens **52 minutes** ago

the bronze age ended about **23 minutes** ago

and each of our lives would pass between **02 heartbeats**

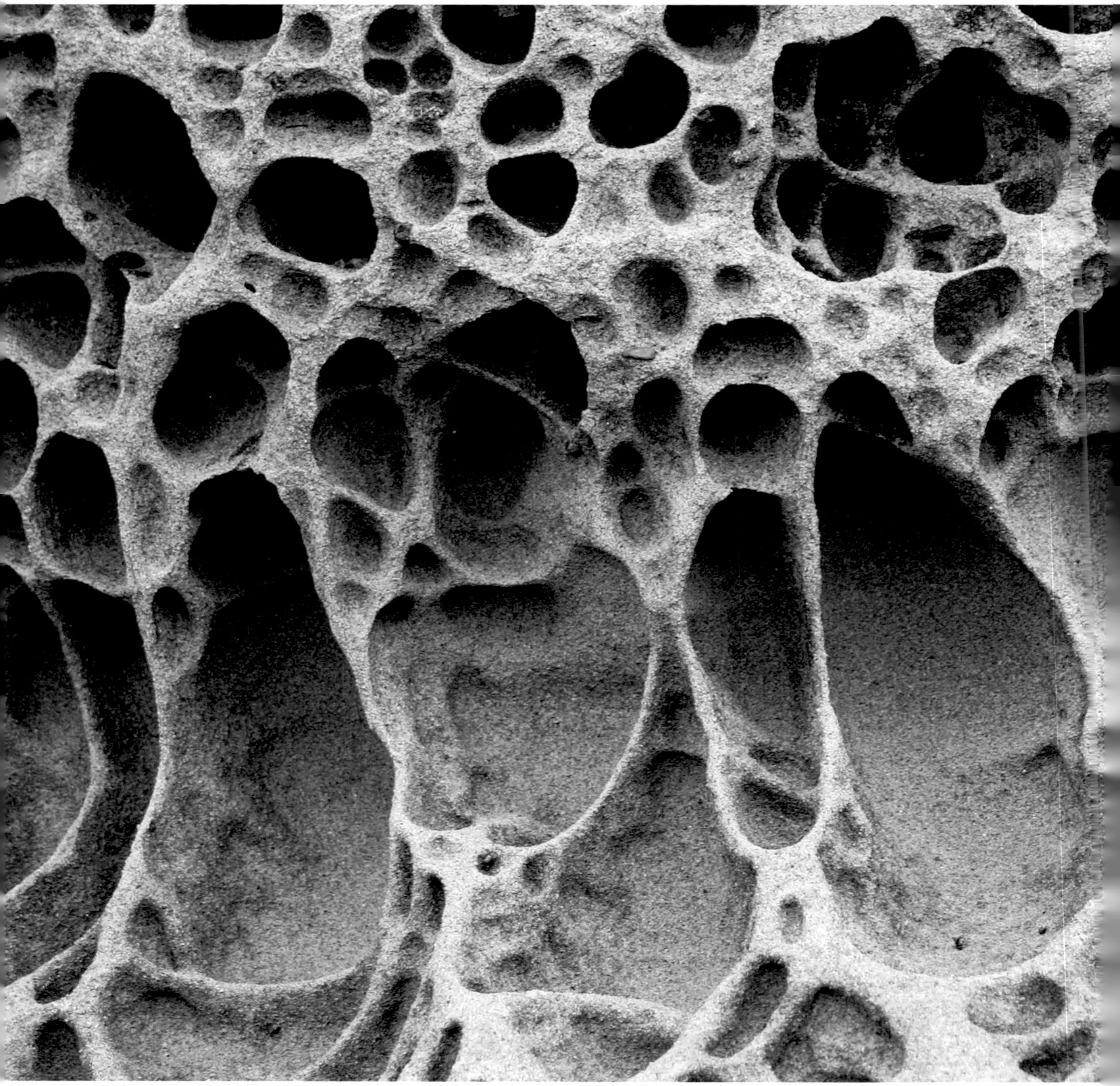

In the uncanny presence

of rock formations we sense

the transience of time

and feel the weight of eternity

contained in the unspoken dialogues

between rocks, the wind and the sea.

Existence as we know it

is possible within the range

of just a few degrees.

This temperature, ideal for life,

means instability for most minerals.

Though imperceptible to us,

rocks are continually changing.

The Earth's inner core releases heat and is thought to influence the movement of molten matter from the outer core, mantle and lithosphere. The inner core is metallic and solid, its temperature 4,000 degrees Celsius. The outer core is molten and fluid, surrounded by a thick mantle that is mostly solid. The lithosphere, on which we live, acts like a strong eggshell that holds all of the internal layers. ¶ The whole surface of the Earth is covered by seven major tectonic plates and half a dozen smaller ones. Tectonic plates are said to be rigid, yet they move like fluid 3-D puzzle pieces over the mantle's molten surface. Continental crusts and oceanic plates float, like islands, over the mantle's surface. Mysteriously, no gaps appear between plate margins although tectonic plates are continually moving; as one pulls away, another immediately fills in the gap. ¶ Plates stretch, push and glide past each other, creating fractures, faults and divisions. Oceanic plates scrape and slip down under lighter, less dense, continental crusts. As ocean plates are pulled under continents, water is released from the sinking or subducting plate. The overlying mantle, or asthenosphere, partially melts and molten matter is released in the form of coastal volcanoes and island arcs. ¶ Dense lava builds up along mid-oceanic ridges and is pushed ahead until it sinks under lighter continental plate margins. Stress between interacting plates results in earthquakes over subduction zones. ¶ The continents, which rest on tectonic plates, change directions, speed by or slow down, and nudge each other along. They push against each other, fold and buckle, create mountains, and merge without being destroyed. Since the Earth formed, tectonic plates have brought all continents together at least four times, Pangaea being the most recent. Rodinia, Columbia and Kenorland appeared as much earlier arrangements of supercontinents.

When in need of stability
we are reminded to
breathe and plant our feet
firmly on the ground.

**Yet, what we call solid
is merely a thin layer of
crystalline rock floating
over molten magma.**

The Earth's underlying forces
help regulate our climate
rearrange and reshape continents
form islands, volcanoes and mountains.

**The bedrock of our existence,
our "stable ground" continually shifts,
ocean floors spread, tectonic plates
move and continents stretch.**

A few hundred years ago

we began to comprehend the

implications and the phenomena

of time as recorded by fossils,

sedimentation and rock formations.

In the early 17th century

the study of rocks and sediments

brought forth a series of new

and challenging questions

regarding the age of the Earth

and the magnitude of time and space.

Creation and evolution theories remain

enigmatic, sometimes conflicting,

always puzzling.

Rocks and fossils provide clues about ancient life. They reflect movement of continents and tectonic plates, contain evidence about mass extinctions and trace evolution. ¶ Pre-biological molecules organized themselves into self-sustaining cycles of chemical reactions that occurred in extremely hot, acidic environments along oceanic ridges. Scientists speculate that phosphates, the spine of DNA, came from meteorites, comets and interplanetary dust, which seeded the surface of the Earth. 10,000 tonnes of extra-terrestrial debris in the form of meteorites still come down every year (curiously, most often in remote areas of the world). ¶ Single-cell, oxygen-releasing, micro-organisms such as blue-green algae and bacteria date back 3.5 billion years. As fossil records point out, multi-cellular life developed 1 billion years later when sufficient oxygen had accumulated in the atmosphere. Micro-organisms flourish and still far outnumber the more complex visible life that began to appear only in the last 540 million years.

Sacred stones
fall from heaven
passing stars
crossing galaxies,
blazing interstellar episodes.
Rocks on the surface
melt with meteorite impact
splash high into the atmosphere
fall back and cool in the shape
of glassy small buttons
small spheres
black teardrops
tektites.

Textures, sensations
absorb desires,
mist and fog
press out sounds.
Warm sun, cool wind and
ocean waves betray;
scurrying organisms enter small
openings – the fluid body of stone.
Tafoni hollows originate
from timeless intertidal
play between water and stone
constantly changing,
teasing us with the mystery and
eloquence of the unknown.

The Gulf Islands of the Georgia Strait, also known as the Salish Sea, contain sediments from rocks that formed under the South Pacific Ocean. About 200 mya (million years ago), ancient North America began to break away from Pangaea, drifting northwest across the Pacific Ocean, collecting exotic terranes and island arcs. Over millions of years, the Pacific Plate intermittently collided with the North American Plate. Independent terranes gradually merged together assembling most of British Columbia's mountain ranges and coastal regions. ¶ 100 mya the Wrangellia Terrane, consisting of exotic sub-terranes collected from south of the equator, tucked its advancing underwater platform into the west coast of North America to become what is now most of Vancouver Island, Alaska and the Queen Charlotte Islands. This proceeding and subsiding platform created the Georgia Basin. Large mountain rivers deposited thick layers of sediments, known as the Nanaimo Group Formations, along the west side of the basin. ¶ 55 mya the Pacific Rim Terrane, and then 7 million years later the Crescent Terrane, collided with the southwest corner of Vancouver Island, compressing the Georgia Basin along a thrust belt under the Nanaimo Group. Sediments over the faults lifted and tilted out of the sea, creating the Gulf Islands. ¶ Much later still, glaciers flowed down the basin and left behind erratics, cobbles, more sediments, and deeply scarred rocks. At the end of the last great ice age, about 12,000 years ago, humans multiplied and many earlier mammals became extinct. Today, each of the Gulf Islands reveals unique sections of stratified rocks that have helped geologists to define the underlying Nanaimo Group Formations.

A vessel in stone

spider webbed relics

tunnels and pathways

unfolding we follow

bone wind, sea swirled, tafoni

– secret pockets of the earth.

Might the density of pebbles and stones vary in direct proportion to their containment of memories and time?

Pebbles, true pilgrims of the Earth

moved by the cycles of the moon

held in the rhythm of the tides

lulled to dream on beaches and coves

nurtured by waves, wind and sun.

Their place of origin disappears.

The Earth's tumultuous inner core
erupts at the seams, vapourized rocks
shoot across the ocean,
clouds of molten rain shower the island.
Waves of intense heat, gas,
steam, ash, and pumice
spread over hundreds of miles.
A blood, red sky envelopes the Earth.
Nine months later,
a single spider spins it's web;
it had blown across the water
in a tightly woven balloon.

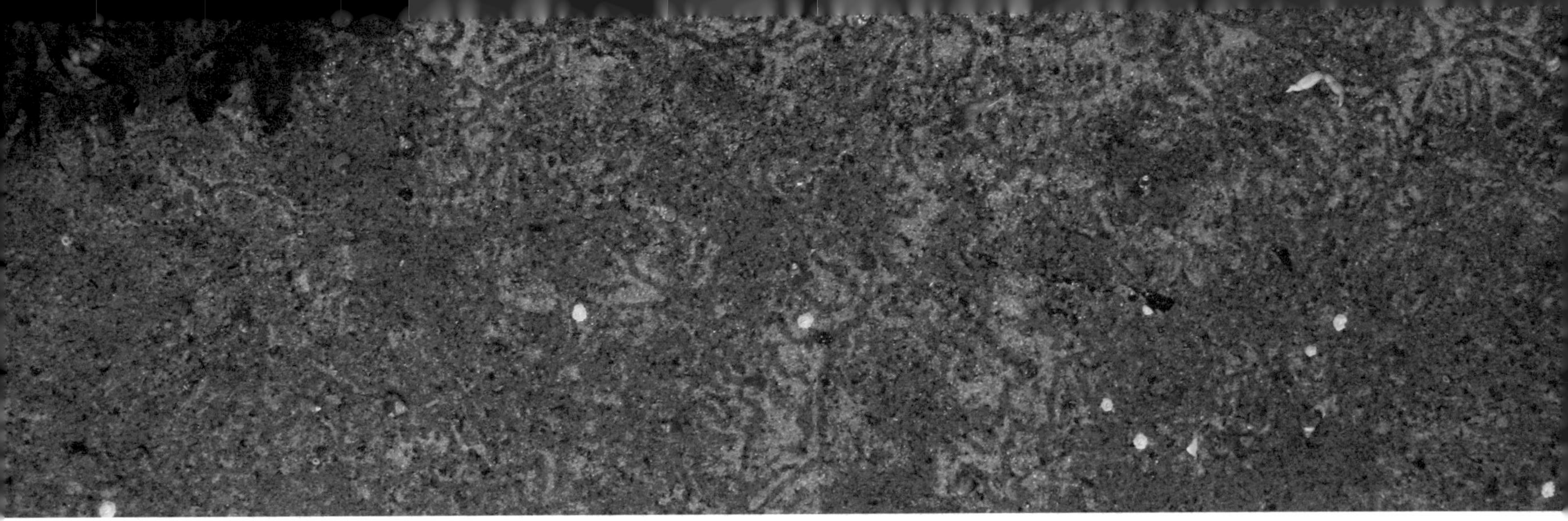

A gelatinous film of new life
initially covers this barren land,
ferns, mosses and flowering plants appear;
seeds brought by birds
scatter across the island.
Butterflies and beetles live on new flora.
Animals swim or drift across the channel
on beach-bound logs.
Over decades, imperceptible changes
leave visceral traces, slowly
the island diversifies and pulses with new life.

rhythm . 02

Underlying igneous and metamorphic rocks of the Georgia Basin extend from Vancouver Island to the Coastal Mountains. Over millions of years, a wide range of sediments accumulated on the basin floor in alternating layers of sandstone, mudstone, siltstone, conglomerates, coal, and shale. ¶ Flying low over the islands we can see that most have long narrow grooves and tilted ridges. Closer to shore, we begin to notice the tafoni hollows that have developed as the softer sediments erode and return to the sea.

The Gulf Islands are defined by eleven sedimentary formations that accumulated over the basin floor. The stratification of each Nanaimo Group is unique. Each formation, several hundreds of meters deep, is composed of distinct layers. Geological events that represent 5 to 8 million years of deposits are recorded in each formation and appear like chapters of an old book begun over 76 million years ago. Each chapter is a series of distinct sediments; some sections are more weathered and exposed than others. ¶ When the Nanaimo Group Formations were being deposited the coastal mountains lay mostly barren. Unobstructed rivers cascaded down mountains and created an extensive watershed of deltas, channels, sandbars, and levees. The sedimentary formations were deposited by a series of large overlapping deltas stretching across most of the Georgia Basin. The majority of sedimentary rocks in which tafoni appear accumulated under a variety of shallow marine conditions that were sometimes swampy for thousands of years and other times grassy, forming overland links. ¶ The movement of tectonic plates, glaciers, sheer weight, and flash floods caused the floor of the basin to periodically sink, then slowly build up again. Erosion from tides and currents left behind sediments in horizontal units of ripples, laminites, silt and coal couplets, turbidities, massive sandstone, and conglomerates. Over thousands of years these sediments hardened in rhythmic patterns and became stratified rock formations. ¶ On the Gulf Islands most tafoni develop on sandstone in the Protection, De Courcy, Geoffrey, Spray, and Gabriola Formations. These honeycomb-like hollows often form between tide lines and at the base of very deep sandstone layers. Larger tafoni caverns also appear higher up on sandstone cliffs resulting from the ocean's spray and wind erosion. Closer to the shoreline, tafoni establish themselves in areas where these layers become irregular, primarily as they transition into harder laminated sediments.

Tafoni and weathered stones

are mnemonic reminders of nature's

eternal rhythms and mysteries.

Sediments fall like unrelenting questions

and ideas that linger and revisit,

until habitual patterns

are turned upside down and recreated.

Long ago, very large coastal rivers created alluvial fans that stretched across most of the Georgia Basin. The Fraser River delta is a much smaller example of such a river. ¶ Deltas accumulate in distinct bed-loads. Braids or bars of inter-connected channels develop and remain exposed, except during heavy rains. Water flows through interloping lenses of gravel and sandbars; mud collects in abandoned channels. Flash floods produce alluvial fans. Torrential rains, confined to a river gully, reach open water and they are forced to slow down. Sediments diffuse, opening like a giant fan with successive layers of imbricated rocks and pebbles. Finer sand and silt fill the crevices in between and the remaining sediments are carried to the edge of the delta, where they form fan fringes. ¶ Fans build outwards and upwards. Heavy rains break through sandbars; new channels and fans open into the sea. Usually, sediments rest as bed loads carried – then dropped – as a current slows down.

The sediments deposited in beds vary from clay particles to fragments, pebbles and cobbles. Most sediments are detritus made of clast, composed of individual grains and fragments. Sedimentary loadcasts later become exposed on sea benches amongst clusters of tafoni. ¶ As a delta advances towards the sea, shorelines prograde: sediments accumulate vertically and horizontally. When the sea floor rises marine conditions become more agitated, creating strong tidal currents. Ripples migrate outwards and finer grains settle during slack tides. Dark mud or black silt accumulate in ripple hollows to form small dark lenses. On flat sandy beaches, as tides alternate, currents and waves create criss-cross patterns of interference; cross-bedding builds up. In deeper water waves move along the ocean surface, while underlying sediments remain relatively undisturbed. When tides recede, tafoni and other undulating forms appear on these layered marine deposits.

Broken light,

the skin of stone

invites us toward

ancient horizons.

Eyes, ears, nose, mouth,

silent compositions

suspended in time.

Restless incidents

composed in rhythmic traces,

while the earth's salty

musky murmurs

escape between its joints.

Our lives are hinged on the measured

flow of cycles, patterns, words, and phrases.

The texture of each language reverberates in

the rhythms of the culture and landscape it contains.

Each family's intonation becomes articulated

by the subtle composition of interbedded DNA.

Cliffs and shorelines reveal stratified sediments that can vary in thickness, from being paper thin up to several meters high. They range in colour from coal to white, with tans, reds and blue greens in between. ¶ Every time the environmental condition of the basin changes, a different type of sediment begins to settle. When more sediments build up than are being transported, sedimentation occurs. Once buried, numerous chemical and physical changes (known as diagenesis) take place before sediments harden. Compaction is the main physical force that forms sedimentary rocks. ¶ Currents produce asymmetric ripples, waves produce symmetric ripples, and both become bedded marine memories. Ripples are made of sand and coarse silt and form under the force of currents or waves. In the sea, currents change with each tide. When a current of water stops or suddenly changes, gravel and pebbles drop and beds have no time to develop. Pebbles imbricate, aligning with the current, their tips point down, face upstream, and they shingle one another. Fine sand and silt settle between the surrounding pebbles. ¶ Turbidity currents occur when sediments slide into channels and the walls give way. The heavy suspended sediments become a forceful current, which slows down and settles as the slope levels out. The load pulls back and climbing ripples form. As the turbidity current slows, finer sediments settle into rippled hollows and form couplets of laminated sandstone and mudstone. ¶ When the Gulf Islands lifted and tilted out of the basin these underlying sediments became the setting for some of the most spectacular and mysterious weathered rocks in the world, which only become visible during fairly low tides.

Rhythm, receptive and giving

continuous internal impulses.

Sensual dialogues interlaced

between crescent moons and

quickening heart beats – discreet,

sweet cadences of life.

Delicate lines drawn aross each face.

Small crevasses, fine folds

caress joys and betrayals.

Between the lips – worlds open.

Beyond the eyes – laughter.

Subtle undulations,

veins rise over tendons.

In silence – touch stone.

Yua is a term used by the Yup'ik people of

Alaska for the spirit that inhabits all things

both animate and inanimate.

A feather has as much yua as a bear,

a cedar tree, a river, or a human being;

each deserves equal care and compassion.

When a Yup'ik goes out for a walk and

sees a stone lying against a tree trunk,

he might pick it up and turn it over.

This would bring a blessing on the stone's yua.

It's hidden side would receive

fresh air, sunlight and rain.

Stately erratics, like proud sentinels on wide sea benches, wait for the ocean's return. Sole marks, or load casts, tell their own stories, as unique sedimentary hieroglyphs. Striated beds of tilted formations appear as if wrenched out of the earth. Dark and light outlines, wavy ridges and grooves, stretch toward the sun. ¶ Most sediments are relocated and transformed many times before they become rock formations. At very low tides waves disturb the sea floor; mud or limestone is replaced by silt and fine sand. Small organisms thrive in these nutrient rich, fresh water currents. Burrowing organisms break ripples apart and when they die, their shells slowly sink and dissolve. Coquina, chalk and selenite are high in calcium and build up from shells and micro-organisms. ¶ Cross-laminated ripples, climbing ripples, normal grading, contact intervals, load-casts, slump traits, concretions and convoluted lamina are distinct traits of the

Georgia Basin's delta fan formations. These kinds of geological deposits appear in small pockets beside tafoni hollows on most of the Gulf Islands. On Hornby Island tafoni form in exposed rock strata, where laminated sediments of silty mudstone and sandstone alternate to create fine lace-like ridges and hollows, carved in thin layers of multi-coloured stone. On the south side of Saturna Island tafoni appear embedded, between laminated joints and vertical fractures, creating magnificent sandstone cliffs. ¶ With climate change ocean levels will either rise or fall. Existing tafoni may become eradicated or radically transformed as their presence is finely tuned with our current environmental conditions. If the ocean levels drop, tafoni can become homes for more lichens, mosses, small organisms, and birds. If the ocean levels rise, tafoni might, for a while, become hiding places for fish and then return as weathered and eroded debris back into the sea.

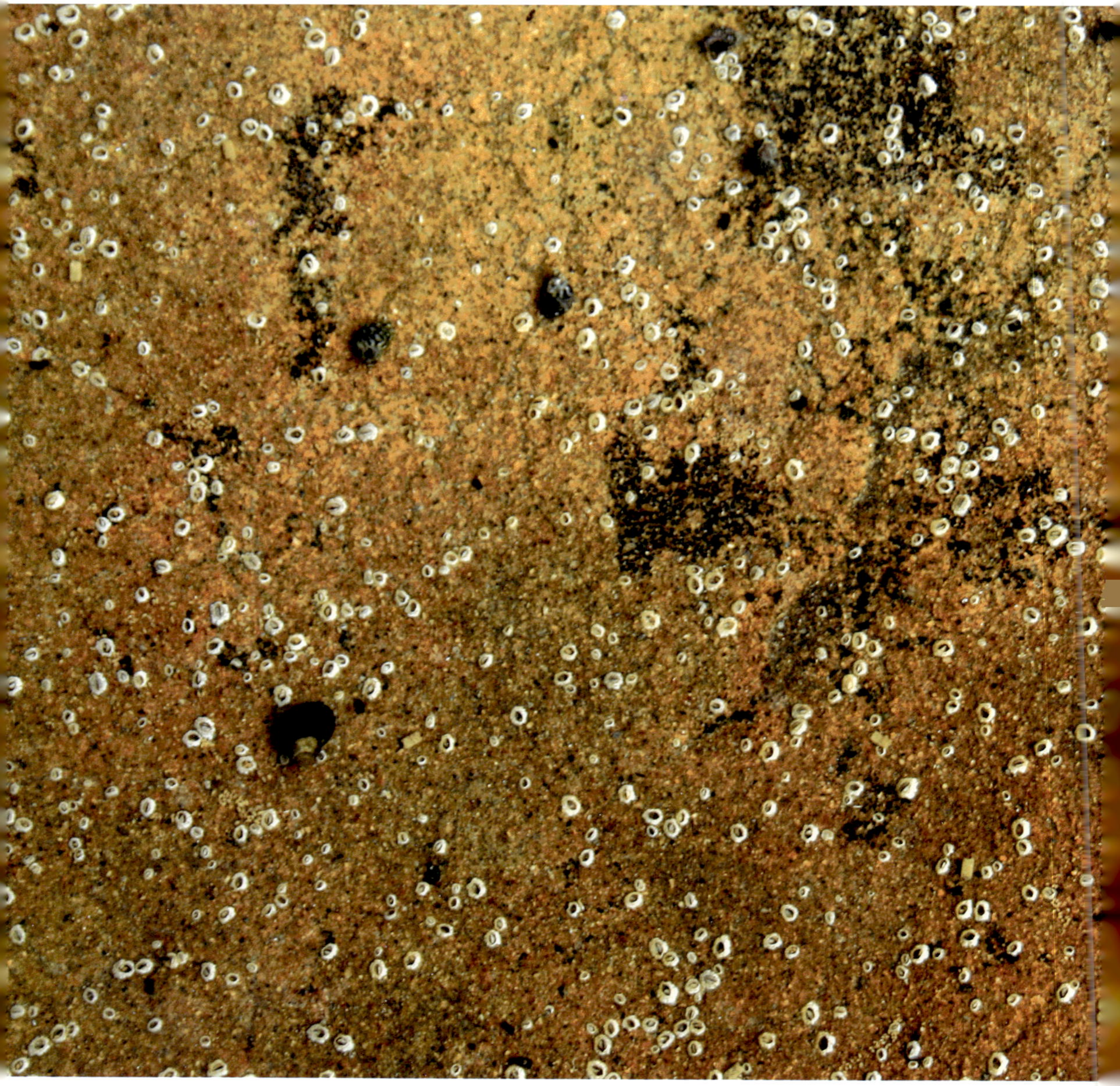

Microscopic ammonites

hovering

white, spiral disks

swimming fossils

in a mellifluent, stone sea.

Over time, our body, like the earth
gathers subtle contours.
Events stitched in tight patterns,
generations embroidered
into the folds of our skin.
Unspoken gestures
stream through the lamellia of
our bones. Autobiographies
pulse through intimate landscapes,
memories ripple into our veins
and heart chambers; iron and salt
slowly settle, undisturbed.
Years later, a pebble drops.
The skin of the ocean breaks,
the tapestry of time unravels,
coded memories unfold
and we tumble
into the fertile sea of
credulous innocence that remains
from our childhood.

erosion . 03

Tafoni weathering, sometimes referred to as alveolar weathering, has recently been of interest to scientists and geologists. With increasing levels of acid rain and carbon dioxide in the atmosphere, many of the world's great monuments and stone buildings are becoming re-ornamented with tafoni's intricate patterns and hollows. ¶ The rocks of the the West Coast are also exposed to various erosive elements that can dramatically change on any given day. Saline moisture, acid rain, high winds, torrential rain, mist, and sun affect the rate of weathering and erosion. ¶ The relatively cool temperatures of the Earth are in part maintained by rock weathering and erosion. A healthy presence of bacteria and algae on rocks and in the ocean not only increase the weathering of most rocks but also sink large amounts of carbon dioxide from the atmosphere back into the ocean.

Between two pebbled coves

a narrow, rocky ledge

stretches towards the sea.

Tonight the moon is pale

rising over a blue-gray horizon

fretting with it's

capricious skirts.

The wind gathers in small pockets.

Between enticing sounds

the tide closes in.

Waves in broken rhythms

hesitate and reverberate

swirling in hollow spaces,

echoes and vacuums,

crystalline chambers,

small glass bells

plucked.

Tafoni strums

transient

eroding

variations

beneath our feet.

Thermal expansion

frost and tree roots

lift and pry open.

Eroded rocks and stones,

the craggy skin of the earth

pulsing, breathing, fluid.

Erosion loosens, dissolves and transports minerals with water, wind, gravity, and ice. Like water's fluid cycle, rocks have a slower, steadier cycle of destruction, transformation and rebuilding. Igneous, sedimentary and metamorphic rocks change during this ongoing cycle. ¶ Igneous rocks come from magma, which is released from the mantle and core of the Earth along oceanic ridges, volcanoes and rifts. Sedimentary rocks form on, or close to, the surface of the Earth. Existing rocks, organisms, boulders, organic sediments, fragments, and pebbles erode, are transported, and weather. They slowly rebuild and solidify under their own weight. Igneous and sedimentary rocks flow with intense heat and pressure beneath the surface of the Earth. The mineral crystals in these rocks realign and often fuse into thin layers, which sometimes fold and reappear as much denser, heavier metamorphic rocks. This uninterrupted cycle has been going on since the Earth appeared; newly formed rocks that come to the surface become transformed again and again. ¶ The weathering and erosion of tafoni require a combination of dynamic climatic conditions, mineralogical properties, topographic features, and other unknown variables to begin. Around the world, they grow in small pockets of sandstone, granite, limestone, dolomite, and metamorphosed conglomerates. Since these diverse kinds of rocks weather differently, tafoni take on slight variations from one location to another. It seems that these hollows become well-established in areas where saline moisture, trace minerals and fairly extreme micro-climates exist. When and where they do form, tafoni are prolific. Local sandstone tafoni seems to have smoother, elongated and more rhythmic patterns than chiselled caverns, which form on granite.

Rocks feel like the stable,
solid ground under our feet even though
their textures, patterns and fissures
reveal signs of dissonance,
change and transformation.
The periphery of each passing wave
imprints silent memoirs;
traced in stone faces are
turbulent seas and tranquil waters.
Never knowing how change will come
we are held in suspension and
always surprised by how quietly it slips in
with every breath that we take.

Rocks sink and rise

rocks tilt and fold

are vulnerable

rocks crack

rocks bend, crinkle, liquefy

break open with earthquakes.

Water erodes

wind weathers away.

The beauty of stones

water

time

light

and trust

are welcome anchors

through uncertainty and change.

Chemical and organic weathering alter coastlines. Carbon dioxide from the atmosphere dissolves in rainwater and reacts with calcium silicate in rocks, widening joints. Clints and grikes appear. Saline moisture also seeps into rocks. Salt expands and crystallizes, producing honeycomb weathering that is, arguably, a different weathering process than tafoni. Iron and oxygen on the surface of rocks oxidize to create a protective glaze or coating of rust, but iron found inside rocks causes them to crumble. ¶ Organic weathering occurs when acids from lichens, bacteria, mosses, and decaying plants destroy minerals in rocks. Air fills gaps, fungi and algae bind particles, leaves protect against the rain, and roots bring minerals to the surface. Burrowing insects, micro-organisms and fungus decompose; their acidic remains convert rocks into dark fertile humus. Plants, trees and small organisms thrive in mineral rich soils and release oxygen. Thus, each breath comes, imperceptibly, from the quietly changing rocks beneath our feet.

Dramas and events

etch delicate patterns into our skin

tuck memories between our joints

and retrace the lives that we have lived.

Time erodes wearied players,

residual tracery hangs where

the ephemera of youth once

entertained our lives.

In the rhythms of change

translucent passions bathe us

with luminosity, beauty and grace.

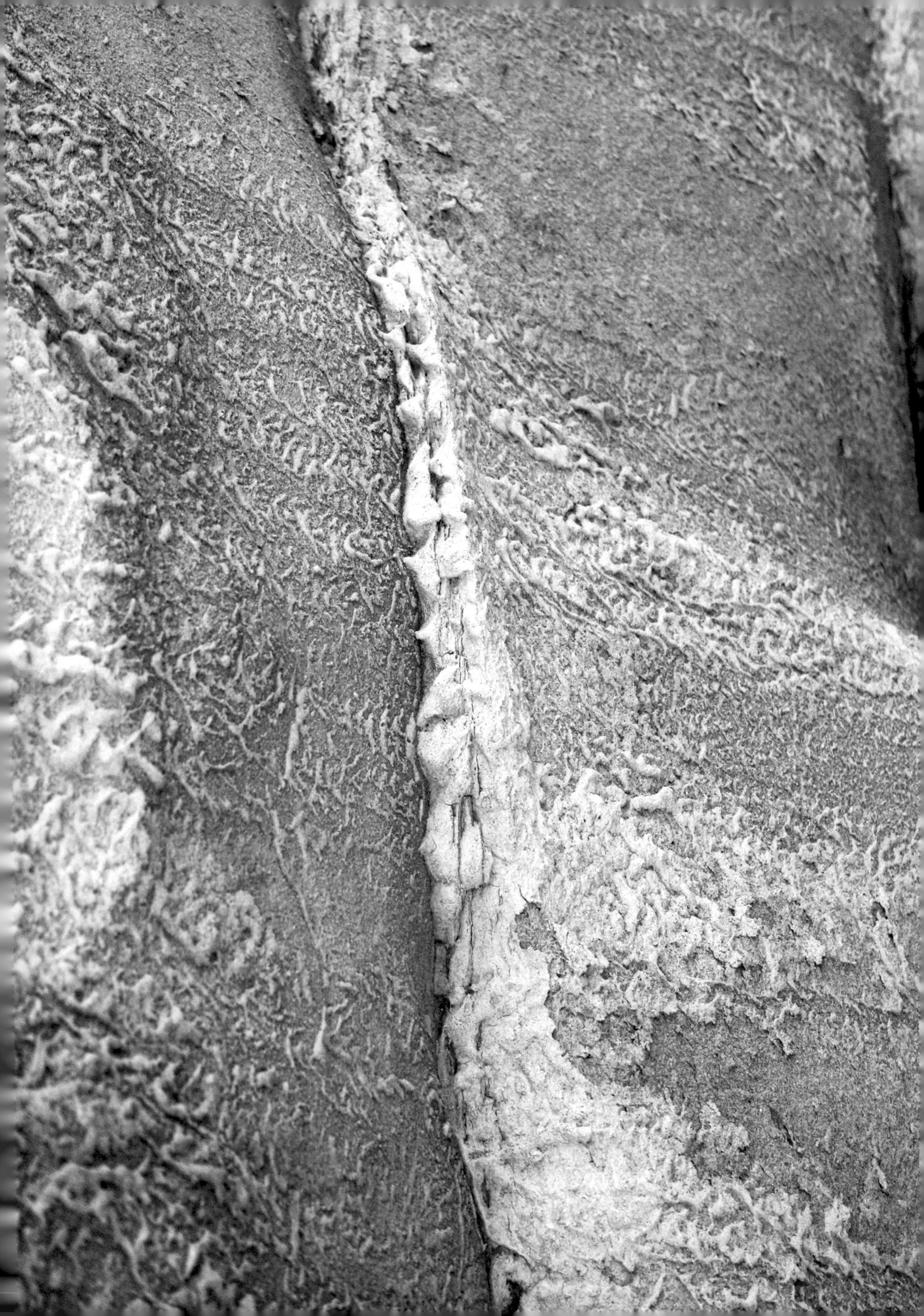

03.100.101 erosion

Waves

expose layers

peel back veils

sensuous forms

hidden bones

and our core remains.

Tafoni can be found on aberrant rocky platforms where saline moisture gathers, pools and retreats. They form primarily on vertical surfaces, close to fractures and hard thin layers, where sedimentary formations transition from one strata or bed into another. ¶ Waves and tides expose hidden parts of rock formations. Variations in climate alter the intensity and level of erosion. Although turbulence is not predictable, wave patterns create rhythms altering the speed and height of currents, which then repeat with the cycles of the moon. Eventually, waves undercut shorelines and large overhead sections collapse. As they carve further into a cliff, wave-cut benches remain, softer sediments erode, and resistant rocks and concretions become outcrops and ridges. ¶ Like intertidal organisms, the majority of tafoni grow between the high and low tide lines. The greatest diversity tends to develop where wave action prevails and varies the most. ¶ Close to the shore and over sea benches waves gather debris, then strike and chisel rocky surfaces. As waves draw back, suction creates cavities and pinholes. Particles and acid rain collect in these tiny pits; waves and the wind spin these irritants, which carve out small circles and spheres. Shallow water platforms create surface tension that triggers a spiraling motion of water molecules around resistant outcrops and mushroom-like pods. Clusters of tafoni and other eroded hollows appear in specific areas of each rock formation, depending upon the micro-environment, mineral composition and wave action. Weathered stones, seaweed and driftwood find shelter between these curious outcrops, delighting children and itinerant beachcombers with unexpected finds.

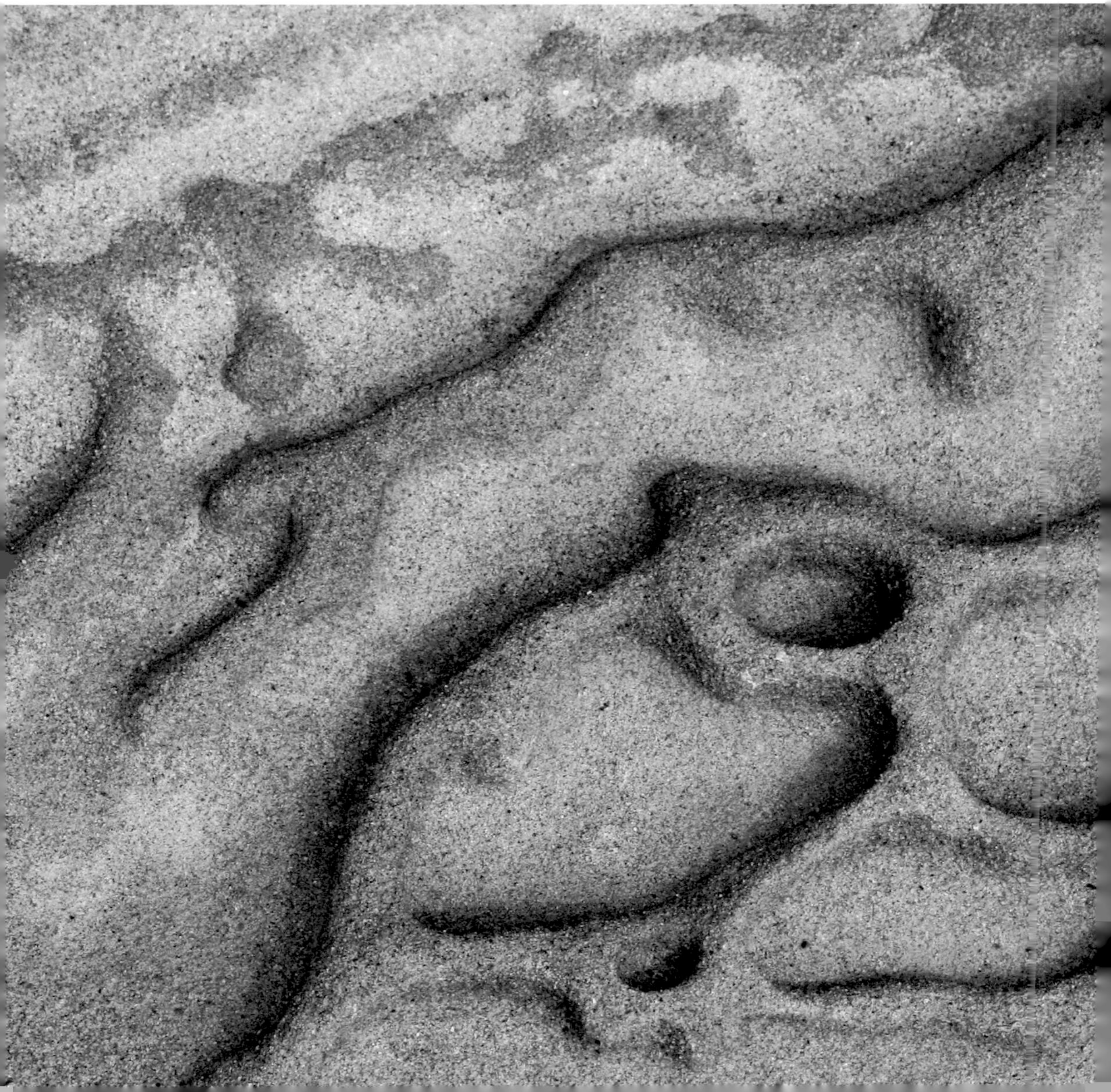

Oceans rise and fall,

tilt our horizons,

alter the rhythm of time

and shift our points of view.

Periwinkles glisten,

long shadows trail

small bundles of light.

Waves linger in hollows,

water becomes stone – stone water

a dance of light and touch – breath

folds over the skin of the earth.

Waves glide across each face to

embrace the solitude of stone.

Water's fluidity and strength

peel away excess layers.

Lingering memories

and deep scars remain.

Our resilient links

to the lives we have lived.

Well-weathered rocks and tafoni formations are not normally studied by geologists – their origin almost impossible to trace and their composition sometimes difficult to determine. The smoother and rounder a stone, the older it tends to be, the longer it has eroded, the further it has travelled. Stones and tafoni are both difficult to date. ¶ There is no general agreement as to how long it takes tafoni to develop. Each tafone hollow unfolds in various stages due to salt crystallization and mineral hydration. Erosion occurs faster once the surface of a rock is broken; each tafone develops an internal micro-climate that is more extreme than its external environment. Each small cavern creates a condition of instability that stimulates and perpetuates its own formation. Stones and tafoni ridges are the last remnants of any given rock formation; both are composed of well-bound particles, crystals and mineral.

Everyone has picked up a pebble or a stone

put it in a pocket

moved it across a beach, across an

then it's forgotten

until we hear it clicking

in the dryer's spin cycle.

03.110.111 erosion

island, across a continent, across an ocean

We surrender, suspended

between strength and vulnerability.

Our flexibility expands and contracts

through the cycles of time

in the exquisite movement

toward the unknown.

In late spring

sea spray hangs

over hillsides and cliffs.

Diaphanous veils

lace the island's

wind swept shorelines.

This morning we tread

down a narrow path

along a wide sea bench

to craggy rock formations

in grapelike clusters,

hollows around stems.

We trace and follow

the resonance of

an eloquent orchestral score.

Listening, we meander

beside the water's edge.

fissures . 04

Forbidden, sensual and sensuous, a fissure is a wonderful metaphor for those in-between places and moments in time of intimate engagements between ourselves and the Earth. Fissures are vulnerable and protected places where small organisms gather and grow. Sometimes microscopic, they are jagged openings that relieve some of the Earth's internal pressure that builds up. Fissures and fractures are the fine lines of darkness and light, where the expansive fluidity of our imagination inquisitively lingers.

How is it that

a broken stone pries

open the heart,

a fissured stone

intrigues with its mystery,

a smooth stone

warms the palm of our hand?

Several of the Earth's outer plates converge along British Columbia's coastline continually moving, stretching, squeezing, creating faults, folds, and deforming rock strata. Faults develop primarily on hard rigid rocks as microscopic fractures or thousand mile-long openings. Faults have dips and strikes, sharp angles relative to the horizon. Folds, on the other hand, vary from a few millimetres to the height of the Himalayan mountains. Folds bend along a hinge; they have an axial plane with an up-throw and a down-throw. Under pressure, formations buckle and rock strata moves like ocean waves. ¶ Faults and folds create irregularities in the composition of rock formations. Pressure and temperature changes cause fractures and openings to appear in areas of weakness. Where the land and ocean meet, flowing water, whether it's in the form of a current, as turbulence, or as a tide, is a powerful force that widens, deepens and smooths out openings; erosion highlights all the irregularities of rock formations.

Roots ease into fractures,

wild rose, salmonberry,

horsetail and mimulus.

Ruby huckleberries drop into

deep, shimmering crevasses.

The earth's vulnerable, raw edges

are protected with wild flowers,

succulents and luminous mosses.

Wave-cut platforms provide abundant places for small organisms – seaweed, algae, sometimes mosses and lichens – to grow. Fractures, fissures and grooves become catchments, not only for these ancient forms of life, but also for tafoni, which establish their roots along these narrow openings. ¶ Changing environmental conditions and underlying tectonic plate movement create areas of compression and fissures in rock formations. With the changing tides, rain and sunshine, hardening minerals dissolve and evaporate to the surface of sandstone. The ongoing movement of water can also precipitate these hardening agents into microscopic fissures. Perhaps these paths of least resistance, where mineral-rich solutions pass, eventually become solidified creating internal tafoni templates. Once the surface breaks, the softer, mineral-deficient pockets erode. Mineral-rich ridges of tafoni emerge where microscopic fissures used to be. ¶ Fractures, grooves and fissures are the vulnerable and dark places that minerals, water, organisms and our curiosity remember and always return to; where things begin, and where they end. Is this what draws our primal imagination to places where it can wander, pose questions and imagine possibilities beyond the what lies on the surface? ¶ The sandstone indentations and ridges of tafoni are like our skin stretched over flesh, tendons, orifices, and protuding bones. They mirror parts of us in uncanny ways, being familiar and awkward at the same time. As we allow our imagination to meander through these ever-changing landscapes we begin to feel the sensuality of the earth breathing under the outlines and shapes that resemble familiar animate forms.

What defines a sensual form,

a sensuous form?

Sensual: sexually appealing.

Sensuous: aesthetically appealing.

Something we desire

to touch, to feel, to sink into,

appealing to the senses,

in two revealing states of mind.

Does the difference lie

in the experience of

how we perceive?

Fissures in weathered stones

expose an innocent eroticism that

is both explicit and intimate.

How is it that the earth

can be so erotic

so sensuous

so human?

The earth's forms

mirror our own sensual bodies.

Or is it that we

are sensual beings,

reflecting her sensuality?

Hidden lives of stones

erotic bodies; sensuality

exposed with each wave.

Vibrant innocence

the breath of the wind

releases crystalline contours.

04.130.131 fissures

how small language is

articulating what the body senses

what the body perceives

what the body knows

Cloaked in vulnerability,

recalling life's transitions,

we enter, being asked to remove

worn-out coats and garments

to uncover our inherent resilience,

sensuality and beauty,

we had so neatly tucked away.

Stone folds into water

creasing along invisible

lines of vulnerability.

Ancient and wildly futuristic,

tafoni's architecture stands

humble and majestic.

Our sense of time, of scale and materials

tumble into unexpected places.

Wounds heal over time

translucent memories

veiled by saline tears.

Smooth elongated hollows

long for touch,

fit our fingers, our hands,

our spines could

curl up in the innocence of

sleep only children know so well.

Sensual patterns and designs

delicate, resilient, intimate

pockets of time

turned inside out.

Future memories

slip into our dreams.

design . 5

Resilience and purpose underlie the beauty and design of the connective tissue in roots, branches, seaweeds, ferns, and lichens. The microscopic patterns of our bones, dentine, blood vessels, cells, and alveoli of our lungs bear a resemblance to tafoni. Could it be, that through timeless, miraculous tinkering, tafoni's intricate cell walls and hollows have become one of Earth's most discernible bridges between geology and biology?

A silvery blue mirror slips over
the ocean's stillness,
ephemeral light reflections are
enticed into intricate hollows.
A child follows outlines
streams, buttons and barnacle pies,
drops pebbles, codes memories,
creates worlds echoed in
playful reverberations.
We pierce through time
and catch a glimpse of
stunningly innocent beauty in
an object, event, place, or person
to realize that these are divine,
poignant moments.

Weathered tafoni and stones embellish Earth's elegant skin. Well-worn, both are composed of resilient materials, which have been brought to the surface, exposed and dramatically transformed. Tafoni are clusters of negative forms defined by surrounding walls; stones are usually solid positive forms. Tafoni and smooth stones are the skeletal remains of eroded rock formations. ¶ Much like mosses and lichens, tafoni develop in clusters or colonies and are said to be self-similar. Their small scale patterns are analogous to their large scale outlines, and like ferns, snowflakes and branches, they create approximate fractals. Their miniature complex shapes reflect the patterns of the greater shoreline they inhabit. ¶ Tafoni erode into loophole patterns, hollows and latticed stone ridges. These complex arrangements appear as fascinating otherworldly architecture, gardens and landscapes.

Like well-crafted work, it is hard to measure how long it takes tafoni to form – between thousands of years to a few months – depending on the host rock and its environment. Tafoni openings can be oblong or slightly irregular circles, and vary in shape, colour and texture from one rock formation to another. ¶ Between tide lines, the ocean erodes the sharp edges of sandstone away. The interior hollows of tafoni break down faster than the outside rock surface. If weathering is taking place, the hollows will be a lighter colour than the exterior surface that becomes hard, glazed or covered in lichens. Tafoni can sometimes mark ancient sea levels, found further back along the shore, or higher up on a cliff. Fluorescent algae, mosses with tiny white spores, and bright orange, grey or black lichens will unassumingly adorn the surface of these cavernous hollows that are not actively eroding.

Each time we return to

an intimate rocky landscape

certain forms, hollows, coves

although always changing, remain familiar.

Through all that we have weathered

these still places are welcome sanctuaries,

secluded harbours, away from the dislocation

of our daily lives. They drop through us like

plumb lines of reference.

Salt water lingers

unravels beauty.

The ocean's ebb and flow

breathes in and out of

cavernous spaces.

Lucidity,

free flowing and vibrant

in children and the elderly

who dwell on thresholds,

weaves through

every moment of their lives.

Gathered in tears

the ocean returns

during times of transition,

births and deaths,

just long enough to

remind us of our own

mortality and essence.

The creative impulse of tafoni exists on many levels, from providing us with design inspiration to exposing a haunting resemblance and relationship to early life. The distinguishing factor between rocks, minerals and life may well be the stardust in our genes. Phosphates form the backbone of RNA and DNA and amino acids produce proteins. Astronomers now speculate that comets and meteorites provided these essential ingredients that made life possible. ¶ Lichens, a very ancient form of life, and tafoni seem to have established a mutual relationship. Lichens generally initiate the erosion of rocks, yet on sandstone, lichens generate oxalates which protect the surface. Lichens also preserve granite in coastal humid climates. Near the high tide line, rocks provide minerals, moisture and stability; lichens cover and protect their surface. Tafoni hollows expand into the rock around sheltered openings. Slow-growing and long-lived, lichens reproduce by means of spores, which disseminate on rock faces. Very primitive and, strictly speaking, not a plant, they are the complex product of a symbiotic partnership between two of the earliest forms of life: algae and fungi. Lichens visually appear and feel like a stone's flower blossoming. ¶ When nature reveals its animated forms, something fascinating and inexplicable begins to happen. Craggy rock formations turn into unexplored territories, and we, their happy discoverers. As with cloud formations, each one of us will see something different. We feel the excitement of finding patterns and designs no one else has seen before. Beautiful designs come out of erosion's work. Faces suddenly appear out of rock formations that intitially look random and chaotic. Cavernous rocks give us the pleasure of momentarily existing between two worlds; the animate and the inanimate come face to face and reflect one another. During such moments, are we more in awe of stones mirroring the animate worlds, or of ourselves, for secretly discovering the animate in these rock formations?

The calligraphy of

life's repeated patterns

gathers around the eyes

radiate innocence.

Leathery hands come to rest

our silent, steady souls

of the heart and

of the earth.

Animal and human sculptures carved into well-weathered rocks and tafoni formations teasingly disappear and reappear transformed: a woman becomes an old man, becomes an eagle. Majestic faces are visible on certain days, under a particular angle of light, only at certain times of the year. Are these playful, animate incarnations purely poetic inclinations? Through our own eyes, the synchronicity of events lends them meaning, yet nature's animistic sculptures surely pre-date our own existence. How would nature explain these unselfconscious evolving faces of itself?

Does the carving of

a face on a rock

become animated

through our own self-recognition?

Was it animated before

we perceived it?

Before light shone upon it?

The lines between order and chance become interwoven. The magnificence of tafoni forms, on a physical level, could be the result of random sedimentation and chance location, accidental erosion and weathering. And yet, is there one thing that has occurred in the last 4.5 billion years that could have been added or left out for the simplicity of a solitary hollow to have formed on a

vast sandstone formation? How many elements must have come together for a rock formation to appear as if growing roots from a giant tree? Nature's willful desire might well be to gently trigger our imagination with tafoni formations, inviting us into a magical innocence and curiosity, engaged in children's devoted play, enchanting these so-called inanimate worlds with wonder.

Drawn by an urgent, steady

pull towards the unknown

we are led to an opening along the

mysterious path of creativity.

What lies hidden?

Time unwinds beauty,

aging welcomes essence,

integrity and honesty.

Resilience grows with erosion,

memories become lithic traces

compressed, folded and wrinkled

under the weight of time.

Tafoni have been a source of design and inspiration for thousands of years. The earliest record we have of tafoni is *The Spring Fresco*, also known as *The Lilies Fresco*. It is one of the oldest known paintings and dates back about 3700 years. The original frescos covered three walls of a small room in a prehistoric Minoan home. What prompted our ancestors to paint sensuous cliffs with joyful swallows and red lilies blowing in the wind, when mythological stories, and religious icons and images were the acceptable forms of representation? The need to enhance our surroundings with images and designs inspired by nature have deep insurgent roots. ¶ Europeans have used these rocky caverns as burial grounds and buildings for thousands of years. Carving out steps, enlarging spaces, adding doors and windows – not to mention beautiful frescos – they become homes, churches and monasteries, cool havens tucked away from the hot sun and bustling city life. ¶ The tafoni formations on the Gulf Islands of British Columbia remain largely unrecognized. Their formation process is still a mystery; tafoni and honeycomb weathering confound scientists and geologists around the world. Geologically speaking, local tafoni form on relatively young sedimentary rocks, which originate from far more ancient, often exotic, terranes. Although large caverns do exist in these stratified rocks, we inevitably return to the miniature, mysterious stone sculptures that transport us into other worlds. ¶ The First Nations people still consider many of the local islands to be sacred. Traces of weathering have left behind meticulous expressions, creating beautiful stone canvases. The Salish people who carved the islands' petroglyphs might also have been inspired by tafoni and eroded stones. Not to vanish, sandstone petroglyphs must periodically be retraced. These sacred myths and memories are impossible to date accurately, as they are etched by hand, stone upon stone, into the shores of the Gulf Islands in the Salish Sea.

From a creative impulse we
tap into a channel of intuition,
that seems more universal than personal,
and precedes conscious knowing.
We live trusting the core
pulse of the unknown,
creating an improvisation of
lines, shapes and textures.
Intuitive senses reach unforeseen
and subliminal places.
Is this the sculptor's
knowing
to tap, right to edge of
the figure veiled in stone?

Ancient weathered rocks and stones

contain precious and essential inspiration.

The Earth's wisdom of design, function

and sensuous exuberance takes place

wasting neither energy nor resources.

still point

This morning,

sheep amble amongst granite boulders

scattered on dew covered meadows.

Selaginella, nemophilia, small orchids,

mimulus, and blue-eyed mary

peer through the wild grasses.

Mosses and lichens

gather in clusters and along delicate lines

on silvery cedar railings.

we wait

Lascivious plumes of mist

shroud tall conifers and reluctantly

pull away from the valleys.

Clouds hang over the ocean – just

beyond arm's reach.

suspended

Silently

between

illuminated

crinkled outlines

drop fine crystalline veils

of light.

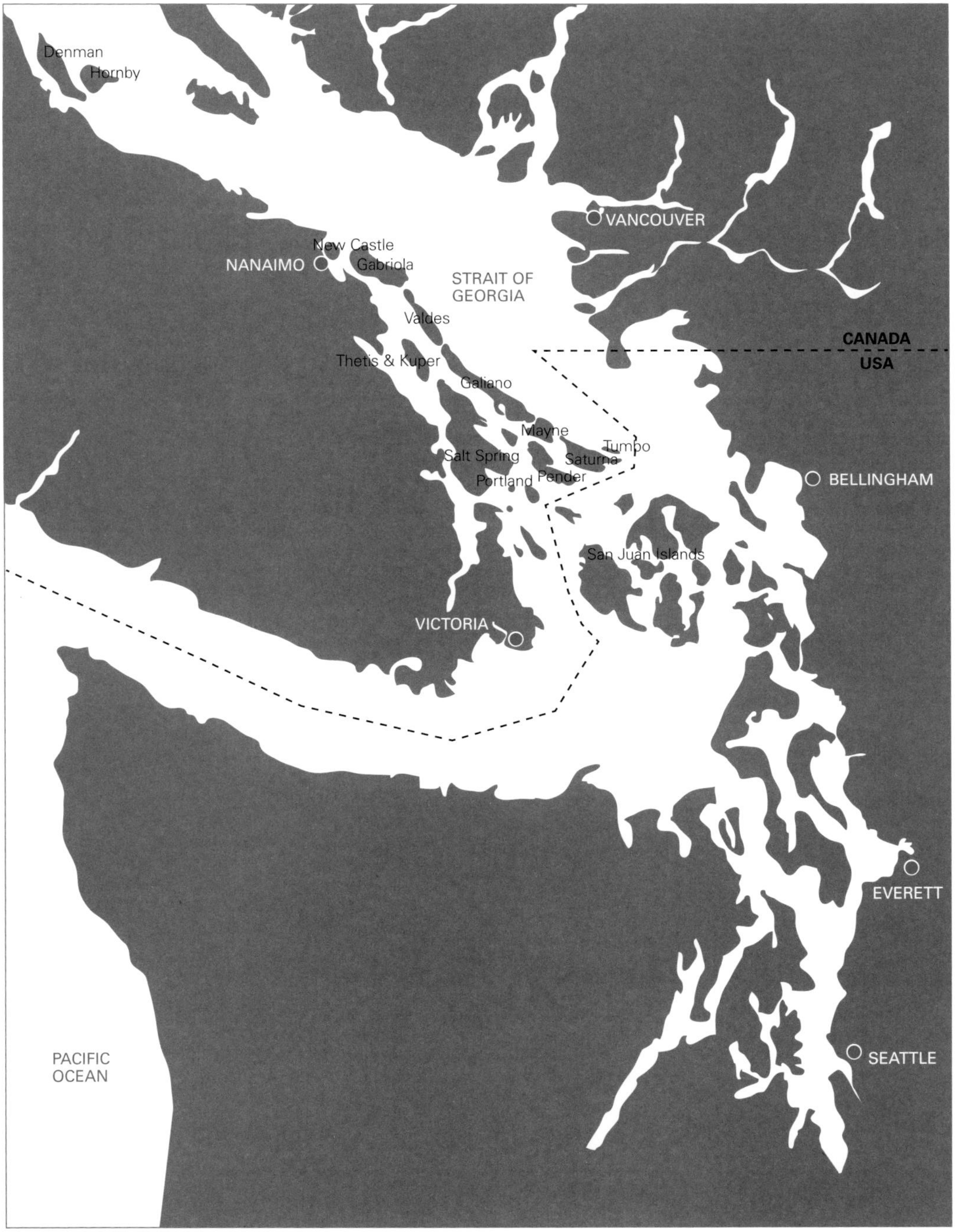
Denman
Hornby
VANCOUVER
New Castle
NANAIMO
Gabriola
STRAIT OF
GEORGIA
Valdes
CANADA
USA
Thetis & Kuper
Galiano
Mayne
Tumbo
Salt Spring
Saturna
Portland
Pender
BELLINGHAM
San Juan Islands
VICTORIA
EVERETT
PACIFIC
OCEAN
SEATTLE

Glossary

Alluvial fan. A sedimentary deposit made by a fast flowing river or stream into an open body of water or broad valley. Sediments drop quickly and spread out into a gently slopping fan shape.

Alveolar. Honeycomb weathering in rocks due to saline moisture and salt crystallization.

Ammonite. A flat chambered and coiled fossil shell of an extinct squid-like creature from the Mesozoic period.

Anticline. An upward convex bend that creates an arch in horizontal rock strata.

Asthenosphere. The top most rock portion of the upper mantle just below the lithosphere. This layer is plastic enough to flow laterally, and varies in depth from 50km to about 400km.

Atmosphere. An invisible film composed of seven layers of gases, which surrounds the Earth.

Bed load. Particles of sand, soil, gravel, and sediments picked up and carried by a current on, or just above, its existing bed.

Braid delta. An alluvial deposit formed at the mouth of a river, usually triangular in shape with excessive sand bars, creating an intricate network of interlaced channels.

Case hardening. Mineral rich solutions evaporate from within a rock; the minerals form a hard coating along the surface.

Cedar District Formation. One of the middle eleven Nanaimo Group Formations, which is approximately 350m. thick and is composed of interbedded mudstone, siltstone, turibidites, shale, and fine grained sandstone.

Channel levee. An embankment or a low ridge deposited by a river, stream or channel of water.

Clast. An individual grain, pebble, fragment, or cobble eroded from a larger mass of rock.

Clint. A projecting hard rock surface or ridge.

Comox Formation. The oldest Nanaimo Group Formation, composed of sandstone, shale, coal, and fossil brachiopods; its total thickness goes up to 650m.

Compaction. The hardening of sediments by bending, fragmenting and rotating grains of rock. At greater depths the points of particles dissolve through pressure solution.

Concretion. A hard round mass of mineral matter found in sandstone; these often build up around an organic nucleus.

Conglomerate. Rounded rocks and pebbles cemented by silicate or clay like minerals.

Contact interval. An interface layer of rock between two distinct rock types.

Convoluted laminae. Tightly undulating rock bedding formed from small scale intricate folds or sub-aqueous slumping.

Core softening. The softening inside of a rock that takes place when minerals in solution are moved out of the core and re-deposited on the surface of the rock.

Coquina. A crumbly limestone composed mostly of fragmented coral or broken mullosk shells.

Couplets. Two alternating layers of sediments create coupling layers of sedimentary rocks.

Cross-bedding. Laminae strata settle at an inclined or traverse pattern relative to the main stratification planes. Migrating dunes or ripples commonly form cross-bedding.

Crystalline Rock. Igneous or metamorphic rock formed of minerals in a crystalline state.

De Courcy Formation. A Nanaimo Group Formation composed of thick bedded medium to coarse-grained massive sandstone deposited in a delta environment.

Detritus. Any loose organic and inorganic material removed from rocks and minerals by mechanical or chemical means.

Diagenesis. The chemical and physical processes involved, that cause a mass of sediments to turn into hard sedimentary rock.

Erosion. The wearing away and transporting of rock material and landforms by the movement of water, wind and rain.

Erratic. Generally, a large rock fragment carried by a glacier or floating ice and re-deposited some distance from its point of origin.

Exotic terrane. A land mass moved some distance from its place of origin journeying on an oceanic plate.

Extension Formation. A Nanaimo Group Formation from the early Campanian period composed of a thick pebble conglomerate sandstone with thin sandstone layers near the top.

Fault. A fracture of the Earth created so that one section to moves, either horizontally and/or vertically, relative to the other side. There are three main types of faults: normal fault, thrust fault and strike-slip fault.

Fissure. A long, deep, narrow cave passage or an extensive crack formed in a rock.

Floodplain. The area of a valley floor adjacent to an alluvial river that is relatively smooth and subject to overflow.

Fluvial. Sediment transported and deposited by a river.

Fold. Rock strata distorted into wave-like forms by pressure.

Foraminiferans. Single-cell marine protozoan whose tiny, limy shells form sediments.

Fractal. A term coined by B. Mandelbrot for complex patterns that Euclidean Geometry could not explain. An object or form composed of small scale patterns that are similar to its larger scale outlines. Clouds, coastlines, snowflakes, ferns, etc.

Fracture. A break, crack or joint caused by stress in a rock; it may also be a break in a mineral along a cleavage plane.

Gabriola Formation. The youngest unit of the Nanaimo Group composed of thick-bedded sandstone and pebble conglomerate and is exposed along the outer edge of most Gulf Islands. This unit is between 466-1000m deep.

Geoffrey Formation. A Nanaimo Group Formation composed of resistant coarse grained sandstone and conglomerate with pebbles, boulders and cobbles.

Granite. A hard and heavy, crystalline, plutonic rock composed of quartz, feldspar and other minerals.

Grike. A solution groove or fissure produced along a crack by the dissolution of limestone.

Grooves. A straight, long, linear depression created with a uniform cross-section and depth.

Groove Cast. A sharp crested, rectilinear or rounded ridge produced when a groove is filled in.

Halite. Common rock salt when it occurs in crystal, granular or compact form.

Haslam Formation. One of the earlier Nanaimo Formations composed of siltstone, shale, mudstone, and sandstone with fish and plant fossils.

Hydrolysis. Water and rock react to turn feldspar into clay or decompose granite into china clay.

Igneous. Molten rocks forced up from the Earth's upper mantle or lower crust. They cool in the ground or on the surface as they come out usually through oceanic ridges, volcanoes and some plate boundaries.

Imbrication. The quick settlement of pebbles out of a current in such a way that they drop in a shingle like formation. The pebbles overlap and align in the same direction; the gaps are filled with fine sediments.

Interbedded. Sedimentary beds with different characteristics that lie between other beds.

Laminite. A sedimentary rock composed of very finely distinguished layers, each being less than 1cm thick.

Lenticle. A rock stratum, bed or body that is lens shaped.

Limestone. A rock composed predominantly of calcium carbonate or magnesium carbonate.

Lithic. An object pertaining to rocks.

Lithosphere. The outer crust of rock on the Earth, about 80km thick, floating over the asthenosphere.

Load Cast. An irregular, bulbous protrusion or hollowed form produced by load deformation, which becomes solidified at the base of an overlying stratum. Usually visible on the horizontal surface of a sea bench where the overlying layer has worn away.

Mantle. A dense, partially semi-molten layer of rock, about 2900km thick, formed between the crust and outer core.

Metamorphic rock. A pre-existing rock transformed by extreme pressure, heat and shear stress, often foliated with aligned mineral crystals.

Monocline. A single dip or slope in a stratigraphic unit of rock.

Mudstone. Fine-grained massive and blocky sedimentary rock formed from mud, 50% each of silt and clay.

Mya. Million years ago.

Nanaimo Group Formation. The name geologists have given to a group of eleven sedimentary formations deposited along the south western section of the Georgia basin. Outcrops of these formations make up the Gulf Islands.

Northumberland Formation. One of the Nanaimo Group formations composed of mudstone, siltstone and sandstone interbeds, contains fossils and is approximately 300m thick.

Ordovician Period. The geological time period when coral reefs, marine plants, and vertebrate first appeared. (490-443 mya)

Paleocene. This is also known as the Age of Ancient Life, ranging from about 543-248 million years ago.

Pangea or Pangaea. The last of the supercontinents formed on Earth, lasting about 100 million years, which drifted apart about 200 mya.

Pender Formation. One of the Nanaimo Group Formations composed of multiple rhythmic beds of sandstone and mudstone with interbeds of fine-grained sandstone.

Polygonal cracking. Joints shaped like polygons, which form and weather on a rock surface.

Prograde. Material deposited by a river delta or fan builds up and eventually moves out towards the sea.

Protection Formation. One of the Nanaimo Group Formations composed of distinct layers of grey to white sandstone, siltstone, carbonaceous mudstone, thin coals, and conglomerates.

Proterozoic eon. The third phase of the Earth's history from 2.5 billion year ago to 540 million years ago after the first super continent formed; by the end of this eon the ozone layer had formed.

Rock Weathering. Carbon dioxide from the atmosphere dissolves in rainwater. It reacts with calcium silicate in rocks making calcium bicarbonate, which sinks into the ocean.

Salish Sea. This has been the new name proposed for the body of water that we most commonly know as the Strait of Georgia. The name change would better reflect the settlement of the First Nations Coast Salish peoples on the coast, which pre-dates European occupation by about 8,000 to 10,000 years.

Sandstone. A sedimentary rock formed from sand particles set in a fine-grained matrix or bound together by chemical cement.

Sedimentary Rock. A stratified rock formed and consolidated over time usually under pressure or chemical process of deposited sediments.

Silicates. The most commonly found crystals in the crust and mantle of the Earth; they contain silicon and oxygen and make up 4/5 of the world's volume.

Siltstone. Silt sediments which have been hardened into sedimentary rock.

Slate. Fine-grained rock derived from siltstone, mudstone or clay like material that breaks into slabs.

Sole Mark. Various hieroglyph-like forms found on the underside of sandstone often formed by organisms, mud and other irregular current deposits.

Spray Formation. The penultimate formation of the Nanaimo Group composed of mudstone, siltstone and sandstone interbeds. This formation is approximately 320m thick and contains concretions and fossils. Its lower portion contains convoluted lamina, ripple cross-laminae and tafoni weathering along the shore.

Subduction. The process that occurs when oceanic plates dive under less dense or less mobile plates, continents or other oceanic plates.

Supercontinent. A large continent composed of all the smaller continents massed together; at least four super-continents have existed and drifted apart.

Syncline. A concave fold created by pressure that forms with younger rocks at its core.

Tectonic Plates. A group of internally rigid, although moving and changing, blocks of the lithosphere, which cover the entire surface of the Earth. They are also known as crustal plates.

Tektites. Glassy objects of debatable origin, believed to have been formed by the impact of meteorites on Earth's surface.

Terranes. Term used to describe parts of the Earth's crust that differ along faults from their neighboring pieces of land; these parts travel independently, riding on oceanic plates.

Turbidite. A rock or sediment transported and deposited by a turbidity current in a graded sequence of grains and particles.

Wave-cut bench. A level platform eroded by wave action; it extends out from a wave-cut cliff. Tafoni form along irregularities and cliffs on such benches along the Gulf Islands.

Wavy bedding. Layers of rock which appear in wavy or ribbon like features.

Wrangellia Terrane. That part of the Earth's crust, which comprises the Insular Belt, Vancouver Island, Queen Charlotte Island and parts of Alaska.

Yup'ik. The Yup'ik speaking Eskimo people live in a diverse range of environments and moderate climates along Western Alaska and Siberia. They live in log and earth homes; they do not live in igloos and have given themselves the name Yup'ik meaning "real person," to clarify stereotypical associations made of Northern peoples.

Further Reading

Abram, David. 1996. *The Spell of the Sensuous.* Toronto: Random House of Canada Ltd. (Language and nature)

Brodo, Irwin, M. and Sylvia Duran Sharnoff, Stephen Sharnoff. 2001. *Lichens of North America.* New Haven and London: Yale University Press.

Butler, Robert. 2003. *The Jade Coast: The Ecology of the Northern Pacific Ocean.* Toronto: Key Porter Books.

Fletcher, Olivia. 1989. *Hammerstone Biography of an Island.* Hornby Island: Apple Press.

Harrington, Sheila, and Judi Stevenson, ed. 2005. *Islands in the Salish Sea: A Community Atlas.* 2nd ed. Surrey: Touch Wood Editions.

Huck, Barbara. 2006. *In Search of Ancient British Columbia.* Winnipeg: Heartland Associates Inc.

Lambert, David, and The Diagram Group. 2007. *The Field Guide to Geology,* New Edition. New York: Checkmark Books.

Lovelock, James. 2006. *The Revenge of Gaia.* London: Penguin Books. (Earth as self-regulating)

Mathews, Bill, and Jim Monger. 2005. *Roadside Geology of Southern British Columbia.* Missoula: Mountain Press Publishing Company.

McBride, Earle and M. Dane Picard. *"Origin of honeycomb and related weathering forms."* Earth Surface Processes and Landforms. Vol. 29, Issue 6, p. 713-735. 2004. Toronto: Wiley and Son's Ltd.

McGraw-Hill. 2003. *Dictionary of Geology & Mineralogy.* 2nd. ed. New York: McGraw-Hill.

Mustard, Peter, S, ed. J.W.H. Monger. 1994. *"The Upper Cretaceous Nanaimo Group, Georgia Basin."* Geological Survey of Canada Bulletin 482. Ottawa: Canada Communication Group Publishing. (p. 27- 95). (Nanaimo Group Formations)

Obee, Bruce. 1992. *The Gulf Islands Explorer.* Reprinted. North Vancouver: Whitecap Books.

Richet, Pascal. 2007. *A Natural History of Time.* Translated by John Venerella. Chicago and London: Chicago University Press.

Rothery, David, A. 2003. *Geology.* Teach Yourself. Chicago: Contemporary Books A Division of McGraw-Hill.

Shorris, Earl. *"The Last Word."* Harper's, August, 2000. (Yua)

Spalding, David and Andrea, Georgina Montgomery, Lawrence Pitt. 1995. *Southern Gulf Islands of British Columbia.* An Altitude SuperGuide. Vancouver: Altitude Publishing.

Stark, Jenny. 1977. *Guide to the Geology of the Gulf Islands.* Vancouver. Department of Geological Sciences, University of British Columbia.

Unione Tipografico - Editrice Torinese. 2000. *Grande Dizionario Della Lingua Italiana,* XX. Italy: Unione Tipografico - Editrice Torinese.

Van der Flier-Keller, Eileen. *A Field Guide to the Identification of Pebbles.* SEOS, University of Victoria.

Yorath, Chris. 2005. *The Geology of Southern Vancouver Island,* Revised Edition. Madeira Park: Harbour Publishing Co. Ltd.

Zimmer, Carl. 2001. *Evolution: The Triumph of an Idea.* New York: Harper Collins Publishers Inc.

On-Line Reading

André, Marie-Françoise and Hall, Kevin. "Honeycomb development on Alexander Island." Geomorphology 65, Issues 1-2, (1 February 2005) <http://www.sciencedirect.com> (5 December, 2007).

Boxerman, Jonathan Z. "The evolutionary cycle of the tafone weathering pattern on sandstone at Bean Hollow Beach, Northern California," Geological Society of America Abstracts with Programs, Vol. 37, No. 4, p. 79, Geomorphology and Quaternary Geology, Session No. 31, <http://gsa.confex.com/gsa/2005CD/finalprogram/abstract_85258.htm> (17 September 2007).

Dorn, R.I. "Case hardening, in Encyclopedia of Geomorphology, ed A.S. Goudie, Routledge: London, 118-119. <http://alliance.la.asu.edu/dorn/CaseHardeningEncyGeomor.pdf> (12 September, 2007).

Earle, Steven. "The Geology and Geological History of Vancouver Island," Department of Geology Malaspina University-College, Canada, <http://webmala.bc.ca/earle/mal-cut/geology-of-vancouver-island. ppt#256,1> (15 August 2007).

Fitzner, B. and Heinrich, S.K. "Photo Atlas of weathering forms on stone monuments," (2004), Geological Institute, RWTH Aachen University, Germany, <http://www.stone.rwth-aachen.de/atlas1.htm> (15 August 2007).

Hacker, Adrienne. Google Search: "Controls of Tafoni Development in Castle Rocks, Idaho," The PDF article provides excellent information on the development of tafoni.

Hejl, Ewald. "A pictorial study of tafoni development from the 2nd millennium BC," Geomorphology 64 (25 January 2005): Science Direct, <www.elsevier.com> (15 September 2007).

Huinink, Pel L. and Kopinga, K. "Simulating the growth of tafoni," Earth Surface Processes and Landforms 29, (2004) 1225, <http://www.phys.tue.nl/MRL/TPM/porous/Restop21.html> (12 September 2007).

Laffineur, R. "Composition and Perspective in Theran Wall-Painting," The Thera Foundation, <http://www.therafoundation.org/articles/art/compositionandperspectiveintheranwallpainting> (23 September 2007).

Marinatos, N. "The Role of Wall-Paintings in the Bronze Age," Art and Religion in Thera, <http://www.therafoundation.org/articles/art/roleofwallpaintingsinbronzeage> (27 August 2007).

Owen, Athena. "A Morphological Comparison of tafoni vs. Flank margin Caves, San Salvador Island Bahamas," 22, October 2006. Geological Society of America Abstracts with Programs, Vol. 38, No. 7, p. 62, <http://gsa.confex.com/gsa/2006AM/finalprogram/abstract_115527.htm> (18 September 2007).

Shelly, C., Robin, M., Plimer, I. R. Editors in Chief. "Weathering," an Encyclopedia of Geology, May 2005, <http://www.sciencedirect.com/science/referenceworks/9780123693969> (8 January 2008).

Turkington, Alice V., Jonathan D. Phillips. "Cavernous weathering, dynamical instability and self-organization," Earth Surface Processes and Landforms, Volume 29, Issue 6, (June 2004) <http://www3.interscience.wiley.com> (20 September 2007).

Additional Resources

Geological Survey is located at 625 Robson Street in Vancouver. This is a government retail store that sells geological maps, geological publications and local interest books about hiking, minerals, paleontology etc. Great resource for everyone.
There is a Geological Reference Library upstairs on the 11th Floor.

The Nanaimo Group Formations

Gabriola Fm.
466-1000m.

Spray Fm.
300m.

Geoffrey Fm.
500m.

Northumberland Fm.
300m.

De Courcy Fm.
300-400m.

Cedar District Fm.
120-500m.

Protection Fm.
30-400m.

Pender Fm.
100-300m.

Extension Fm.
45-500m.

Haslam Fm.
50-500m.

Comox Fm.
200-650m.

Wrangellia Terrane
below

Tafoni on the Gulf Islands

G A B R I O L A
Drumbeg Provincial Park – Gabriola Formation.
Leboeuf Bay – north east shore of the island.
Malaspina Galleries – Gabriola Fm. - great for kayakers.
Petroglyphs – behind the United Church on South Road.

G A L I A N O
Dionisio Point – Gabriola Fm.
Bellhouse Park – Gabriola Fm.
Coon Bay – Gabriola Fm.

H O R N B Y
Hornby Island has large quantities of tafoni. When the tide is right it is possible to walk around the whole island in a day. Tribune Bay – Spray Formation around the south side of the bay; Gabriola Fm. along the north side of the bay out to Helliwell Park. Helliwell Park – Gabriola Fm. High cliffs with tafoni below accessible only at lower tides. Ford Cove – Northumberland Formation out to Heron Rocks – DeCourcy Formation. The Northumberland and Geoffrey Formations are exposed along the North and East shorelines of the island.

M A Y N E
Oyster Bay – Gabriola Fm.
David Cove – Gabriola Fm.

N E W C A S T L E
Protection Formation.

P O R T L A N D
Northern end of the island has beautiful tafoni. The island is accessible by boat taxi from Sidney.

S A L T S P R I N G
Ruckle Park – small patches of tafoni can be seen on sandstone near the ocean. The south end parks on Salt Spring Island have relatively older granite covered shorelines with the beginnings of tafoni on them. Granite is a much harder rock than sandstone. The Gulf Islands are geologically very young and tafoni is slower to form on granite. Cusheon Cove and Long Harbour also have tafoni.

S A T U R N A
East Point Park, Fiddler's Cove, Echo Bay - Gabriola Fm.
Exceptional formations of tafoni and other unique erosions.

T U M B O
Beautiful large formations on the south side – Gabriola Fm.

Tafoni formations are found thoughout the Strait of Georgia.

Every effort has been made to provide acknowledgements and accurate information. There has been little, if any, specific research done to date about tafoni on the Gulf Islands. We will be pleased to correct any unintentional omissions in future editions of the book.

ISBN NO. 978-0-9809951-0-7
Library and Archives Canada Cataloguing in Publication
Viviani, Cristina I., 1962-
Tafoni: natural design of weathered stone / Cristina I. Viviani, author;
Maureen Welton, creative director; Kenneth Torrance, design.
Poems and photographs.

Includes bibliographical references.
1. Weathering–Poetry. 2. Rocks–Poetry. 3. Nature (Aesthetics)–Poetry.
4. Weathering–British Columbia–Gulf Islands. I. Welton, Maureen, 1960-
II. Torrance, Kenneth, 1980- III. Title.
PS8643.I93T33 2008 C811'.6 C2008-901934-2

Published 2008 by:
18KARAT Publishers Inc.
A division of
18Karat International
Product Sourcing Inc.
5292 - 272nd Street
Langley, BC
Canada V4W 1S3

Author: Cristina I. Viviani
Creative Director: Maureen Welton
Design: Kenneth Torrance
Photography: Cristina I. Viviani, Maureen Welton, Kenneth Torrance

The tafoni inspired home décor
collection can be seen at:
18KARAT Home Store
3039 Granville Street
Vancouver, BC
Canada V6H 3J9
www.eighteenkarat.com

Printed in Canada
Special Thanks to:
Raul Vicencio, Peter Cardew, Laura Leach,
Elizabeth Mancini, Robert Sturrey,
Don Simpson, Joan Van Paridon, Sharon Romero,
David Lamb, Rick Findlater, Total Graphics,
Tracy Clark, Hemlock Printers,
Glenn Welton, Grant Ohman, the staff at 18Karat,
Meaghan McBride, Vertigo West,
the staff at the Geological Survey of Canada and
the Instituto Italiano di Cultura in Vancouver.

Mixed Sources
Product group from well-managed forests, controlled sources and recycled wood or fiber
FSC www.fsc.org Cert no. SW-COC-1563
© 1996 Forest Stewardship Council
10%

This book is printed on paper that is made from 10% post-consumer recycled fibre. The coated text pages are FSC certified. By choosing FSC certified paper, 18 Karat is supporting the growth of responsible forest management.

Cristina Viviani was born in Chile and has studied in Europe and Japan. She has been living in Vancouver since the early 1990's. Cristina completed a Bachelor of Architecture and a Master of Fine Arts. The mysterious, the unseen, and often forgotten inter-connections that we have with the natural world have been the source of inspiration for her interdisciplinary work. The wonder, beauty and quiet intelligence of the earth nurture her creative spirit.

Maureen Welton is the President and Creative Director of 18Karat, a Vancouver, BC company that develops and imports home décor products from around the world. She brings the creative vision to the company and is responsible for the product collections and catalogues. Her first enterprise was a retail florist and her connection to organic forms, colours and textures clearly show in her work. She has an enduring fascination with design on all levels, in all forms.

Kenneth Torrance is a graphic design graduate of Emily Carr Institute of Art + Design in Vancouver, BC. He is a stone carver, a painter and a craftsman using natural materials, forms and concepts. Kenneth's extensive travels and encounters in nature keep him connected to his creative source. He has worked as a Graphic Designer in London, England and now works for 18Karat.